Federico Colás Marín

Make biodiesel at home
"Backyard biofuel"

Design: Federico Colás Marín.
First edition: september, 2012
Translation: Paola Martínez
© 2012, Federico Colás Marín

Printed in CreateSpace, a DBA of On-Demand Publishing, LLC

ISBN-13: 978-1478290148
ISBN-10: 1478290145

To Soledad, my support all this years.

Introduction

Dear reader, first of all, I would like to thank you for purchasing this brief but valuable manual about the production of biodiesel.

As you might imagine and expect, the information presented here is intended to assist you to make your own fuel. You will no longer rely on oil companies, and you will enjoy the process.

Even though you cannot create your biodiesel manufacturing company with this manual, I hope you can make a few liters of fuel in the garden of your house to feed this gadget with wheels that you have parked at the door.

I will introduce you the procedures of making biodiesel step by step. In addition, I will talk about some details that I have in mind when producing the precious and unknown biodiesel.

You will also learn some chemistry basics, so you won't be overwhelmed while working on the fabrication, which will help you to understand the processes that take place inside of the reactor, and you'll see that it's easier to reach good fruition with this knowledge. In any case, it will serve you to impress people at some chemists' party.

You can take a look at http://www.biodieselcasero.com where you can find additional information. Do not forget to visit it.

Needless to say, all the information presented here is only for educational purposes and does not constitute a contract, nor make me responsible for the use you give to this information or for any damage you cause to yourself, third parties or things.

It is also your responsibility to comply with the laws applicable in the territory where you live. Learn about them as it might be illegal to produce biodiesel in your country. If this is your case, I highly recommend you to stop reading.

With the above information explained, I think we can proceed.

What is biodiesel?

Biodiesel is a liquid obtained after a transesterification (in methyl esters) process of existing fatty acids in vegetable oils (soybean, sunflower or canola, for example). Although its characteristics are similar to the diesel fuel, we should say that biodiesel has a flash point considerably higher than diesel fuel, which makes it safer.

In regards to the diesel fuel:

- Biodiesel is more biodegradable; thus, discharges into the soil do not cause irreparable damage to the ecosystem.
- It is more detergent, and it takes off all the deposits left by the diesel fuel and, once the internal circuit of the engine is clean, it keeps the engine as new.
- Their combustion gases do not have high concentrations of sulfur dioxide, particles, heavy metals and volatile organic compounds.

In regards to your finances:

- You will not depend on the whims of an oil company. The price changes in the raw materials that you will use are lower.
- The cost of producing a liter of biodiesel is lower than the price per liter of diesel fuel.
- You will save lots of money, approximately the value of some vacations. It is very motivating to know that the vacations were paid off for themselves.

Remember, you will get fuel from waste. What could be better than saving money, being green and making the exhaust pipe to smell good?

If you have a mill, an olive grove or a sunflower plantation, and you have extra oil that does not pass quality control standards for human consumption, these procedures may serve you to give a usefulness of that remaining production.

In this book, we will focus on making our fuel from used frying oil because using new oil purchased at the supermarket is not economically cost-effective.

What do I need to make my own biofuel?

Initially, you will need very little; however, it is recommendable to make a list:

> 1) Be knowledgeable about basic chemistry. Relax, everything is included here.
> 2) Used vegetable oil, used car oil does not count.
> 3) Some liters of alcohol, usually methanol.
> 4) Sodium hydroxide, also called lye.
> 5) A reactor where to put all the items listed above. We will give you some ideas here.

I am sure I forgot to mention other items; nevertheless, if you are reading this manual without skipping anything, there will be no problems in the process.

Does it require much investment to start?

No, absolutely not. Usually, with the savings made in the course of a year, you will recover both the cost of this book and the cost of materials. If you get to the end and do things right, you can save a lot of money.

This is a long pathway, the vacations that we were talking about on the previous pages may take a little while to come; however, if you take this work as a hobby, and you pay close attention to this manual, they will come eventually.

It is important not to save in safety. People's lives and the integrity of their belongings are more important than biodiesel and all the savings you want to get, so please do not think it twice. Go to a store of work safety equipment and let yourself be advised by the specialists that will assist you there.

It is also important the legal aspect of making biodiesel. The fact that it is legal to purchase this book in your country does not mean that making biodiesel is legal too. Find out whether you can make and use biodiesel in your automobile. Logically, in the event of not being legal, I ask you not to proceed any further in this process, and if you do, IT WILL BE AT YOUR OWN RISK.

A little bit of chemistry

We will begin with a brief chemistry class to know about the process that we will be performing.

Chemistry is very similar to cooking. You can follow a recipe and obtain satisfactory results, but you will never be a good cook if you do not know the characteristics of each of the ingredients that you use, if you do not know what ingredients you should not mix or if you skip the control of basic parameters such as quantities or temperature.

We will see the structure of different molecules. These molecules are composed by atoms that, according to their order and type, define their chemical and physical characteristics. Knowing this structure allows you to predict the behavior of the reagents that you will work with, and you could use this behavior to obtain the products that you desire.

In the following chapters you will also know the chemical reactions that occur between these molecules, and how to manage them. In addition, you will see that those reactions, regardless of their reagents, can be taken to wherever you want if you know what laws define and conduct them.

The information I present here is very simplified, and I will try to provide you with daily life examples that will serve you to relate them with things that you are surrounded by.

We are surrounded by chemistry and reactions that occur around us; therefore, you should not be afraid to experiment. Losing the fear of experiencing does not make you an unconscious person, but you must investigate the theory and, when you master it and know all the risks, you can begin to practice with the security of not wasting time doing tests, nor lose something on a fire, indeed.

Let's proceed...

Hydrocarbons

Hydrocarbons are molecules composed by carbon (C) and hydrogen (H) atoms. These molecules have names such as methane, ethane, propane, butane, etc. depending on the number of carbon atoms that its structure has.

These molecules are the main components of the petroleum that you will stop using soon to feed your car, and the gas that you use for cooking. Organic solvents are also hydrocarbons. Plastic bags, pipes and even possibly the sole of your shoes are made from hydrocarbons.

In Table 1, you will see some hydrocarbons with their name and structure.

Name	Structure
Methane	CH_4
Ethane	H_3C — CH_3
Propane	
Butane	
Pentane	
Octane	

Table 1 – Hydrocarbons

Low-molecular weight hydrocarbons (few carbon atoms) are usually gaseous. As the number of carbon atoms increases, they become liquid and solid at room temperature.

In general, due to their electronic characteristics, they are not very soluble in water. For example, if you mix water and gasoline, they will end up separating from each other eventually.

These hydrocarbons can have replaced one of their hydrogen (-H) for a hydroxyl group (-OH); thus, we obtain an alcohol. The name of this alcohol will be derived from the name of the hydrocarbon ending in -ol.

Alcohols

Alcohols are hydrocarbons that have linked a hydroxyl group (-OH) to any carbon atom (C). If in the molecule there is only one hydroxyl group, we are talking about a monoalcohol; if there are two molecules, it is a glycol; and if there are three molecules, it is a triol.

They have the structure:

R-OH

Where R is an alkyl chain (a hydrocarbon) and -OH is the hydroxyl group. The hydroxyl group (-OH) replaces one or more hydrogens (-H). In table 2, you will see some alcohols with their name and structure.

Name	Structure
Methanol	H_3C—OH
Ethanol	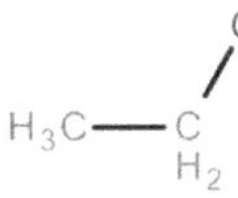 H_3C—C (with H_2 below, OH above)
2-Propanol	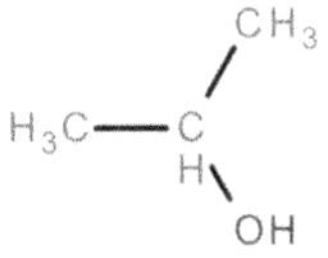 H_3C—C (with CH_3 above, H and OH below)
Ethyleneglycol	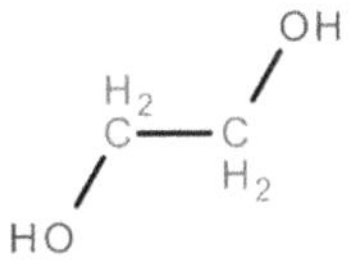 HO—CH_2—CH_2—OH
Glycerol	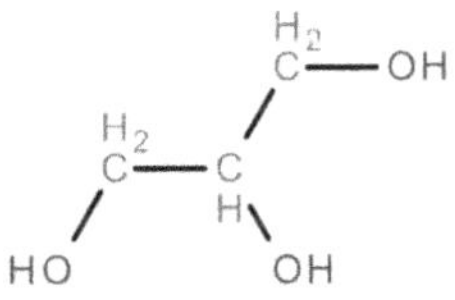 HO—CH_2—CH(OH)—CH_2—OH

Table 2 – Alcohols

Possibly, these names may sound familiar to you:

- Methanol, is the alcohol used for burning purposes.
- Ethanol, is the alcohol used in alcoholic beverages.
- 2-propanol, is used as a cleaner agent.
- Glycerol, is glycerin soap.

It is the hydroxyl group that makes most of the alcohols to have a better water miscibility than the hydrocarbons. The presence of the oxygen atom (O) allows water to be combined with greater ease.

This combining process by which liquids and solids are dissolved is called solvation. The solvation is responsible for making the soap to clean as it does. There is a lot of literature available, go to your nearest university library.

Acids

The organic acids that you will see here have the structure:

R-COOH

Where R is an alkyl chain (a hydrocarbon).

Organic acids can be obtained by oxidation of alcohols. The most common example is vinegar (acetic acid) originated from the oxidation of ethanol made from wine.

We will distinguish three types of acids:

- Minerals (sulfuric acid, hydrochloric acid, etc.)
- Organics (acetic acid, ascorbic acid, etc.)
- Fatty - which are organic acids with more than 14 carbon atoms - (stearic acid, palmitic acid, etc.)

The members of each of these categories have a series of characteristics that will serve us for certain purposes. We will use the mineral acids as accelerators of some reactions, and the organic acids can be used as correctors of certain reaction conditions.

Fatty acids, on the other hand, can be a problem because they are one of the ingredients needed for the formation of soap (which can annoy us a lot) and can also be part of the solution if we turn them into biodiesel.

It is important to know what we can do with the materials that we work with at all times; hence, you must know them.

In Table 3, you will see some acid with their name and structure.

Name	Structure
Formic acid	
Acetic acid	
Sulfuric acid (an inorganic acid)	
Stearic acid	
Palmitic acid	

Table 3 – Acids

These acids are very close to us:

- Formic acid is the acid that ants inject when biting.
- Acetic acid is the main component of vinegar.
- Sulfuric acid, can be found inside of the car battery.
- Stearic and palmitic acid are present in fats.

You will be pleased to know that fatty acids are not soluble in water; but they form soaps in the presence of water and sodium hydroxide (NaOH). Soaps are not exactly friends of the producer of biodiesel. You will find them sooner or later.

Fatty acid salts

Reacting a fatty acid with a base, for example the sodium hydroxide (NaOH), forms a salt which is sometimes referred to as soap. These salts have the particularity of being able to "combine" two liquids which in principle should be separated as the water and oil would.

Fatty acid salts are very large molecules with two distinct areas from one another: one is water-soluble and the other one is fat-soluble.

If we react stearic acid with sodium hydroxide in aqueous medium, we will obtain sodium stearate.

Stearic acid

Sodium stearate   Hydrophilic zone

The area called hydrophilic (which is similar to water), will be soluble in water while the rest of the molecule (being essentially a hydrocarbon) will be soluble in fats.
This will cause that, when we combine water and fat in the presence of these salts, they won't be separated in two differentiated phases. The soap formed will keep every drop of fat attached with other water molecules and these ones to other fat molecules. This is how it forms what we call emulsion.

The soap that you use to wash dishes, to take a shower, and to wash the car are very similar compounds of sodium stearate. All of them have an area which is dissolved in water and another area which is dissolved in fat.

Breaking the emulsions is one of the topics that generate more doubts to biodiesel producers; though, it is not particularly difficult if you know the reason of why they occur.

Esters

Esters are organic compounds in which an alkyl group (hydrocarbon) replaces a hydrogen atom (-H) in an organic acid.

They have the structure:

R1-COO-R2

Where R1 and R2 are hydrocarbon chains.

The most common esters in nature are fats, which are esters of glycerol with fatty acids (stearic, oleic, etc.).

These compounds are more hydrophobic (water repelling) than acids and alcohols; therefore, they tend not to be mixed with water or with low-molecular-weight alcohols and acids.

Biodiesel is a mixture of methyl esters of fatty acids, in few words, they are esters formed from fatty acids present in the oil and methanol.

In Table 4, you will see some esters with their name and structure.

Name	Structure
Butyl acetate	
Allyl caproate	
Methyl oleate	

Table 4 – Esters

A little bit of chemistry II

We already know about some chemical compounds. I will not continue detailing more since the purpose of this book is not that you obtain a degree in chemistry. We do not need more competition. We just want you to make your own biofuel. So, let's move forward with reactions:

Esterification reaction

Esterification reactions have one ester as a final product from the dehydration of an alcohol and an acid. This reaction takes place in an acid medium.

In general:

$$Acid + Alcohol \rightarrow Ester + Water$$

For example:

The octanoic acid reacts with methanol to produce methyl octanoate.

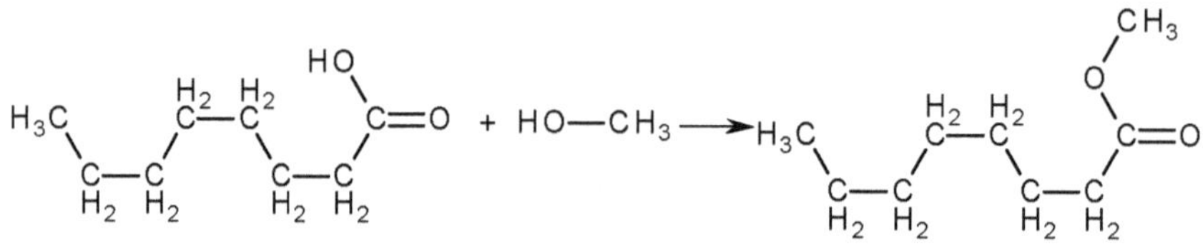

This reaction is performed in the presence of an acid (which I have not written in the reaction) and also generates a water molecule per molecule of ester produced.

If, instead of using octanoic acid, you use a fatty acid, oleic acid for example, you will obtain a methyl ester called methyl oleate that can be considered as biodiesel.

Through this reaction are obtained compounds as useful as the acetate cellulose, aspirin, or benzocaine.

As I said a few pages ago, these reactions are all around you, and you must not feel fear of them.

Transesterification reaction

In transesterification reactions, an ester and an alcohol react, resulting in a new ester that contains a different alkoxy group (R-O- from the alcohol reagent) and an alcohol molecule from the outgoing alkoxy group (which was part of the original ester). These reactions can be performed in acidic or basic medium.

In general:

$$R1\text{-}COO\text{-}R2 + R3\text{-}OH \rightarrow R1\text{-}COO\text{-}R3 + R2\text{-}OH$$

For example:

The propyl octanoate reacts with the methanol in the presence of a base (NaOH) to obtain methyl octanoate and propanol.

These reactions are widely used in the food industry to change the properties of certain fats and thus, avoid crystals that may damage the aesthetics of the products.

One of the problems that can occur in this type of reactions and in reality in all reactions, is that the reactants get separated and then, we are obliged to maintain a relatively intense agitation in a way that form small droplets of a reactant within the other.

The more agitation, the greater will be the amount of these droplets and therefore, greater the amount of reaction zones. The more reaction zones, the better will be the performance and speed of the reaction.

This is very important in the production of biodiesel, in which we will use the transesterification as main reaction, because the methanol and the oil (reagents that we will use) tend to be separated.

Equilibrium reactions

I will try to explain this in an easy and possibly erroneous manner, so I apologize to chemical colleagues who read this and find that I have ommitted a lot of things along the way to simplify the idea.

When we perform a reaction between two or more reagents to obtain one or more products, something occurs that we called equilibrium. That is, when A reacts with B to give C and D, the final reaction concentrations of A and B are not 0.

In a reaction:

$$a\,A + b\,B \rightarrow c\,C + d\,D$$

At the beginning of the reaction, we will have a certain amount of A and B, no C or D. As the reaction progresses, the amounts of A and B decrease while the amounts of C and D increase.

At the end of the reaction, it is expected that the amounts of A and B to be 0 since they became into C and D, but this is not the case. You will obtain very little of A and B and a lot of C and D, but you will always have a bit of the reagents.

The point at which the quantities of C and D stop increasing to remain constant is called dynamic equilibrium. At that moment, it is verified that the relationship between the concentrations of A and B compared to those of C and D is constant.

We could write:

$$constant = \frac{[C]^c [D]^d}{[A]^a [B]^b}$$

If we want to get a little more of the compounds C and D, to maintain constant relationship, we have to increase the quantities of A and/or B. Also, we could decrease the amount of C or D in any way.

This is called "shift to the right reaction."

We will use this "trick" to obtain better performance in our reaction of production of biodiesel. We will remove one of the products, so the reaction will be forced to continue producing to reach equilibrium.

Catalyst

A catalyst is a compound that is capable of accelerate a reaction (or make it slower) by being present in a reaction and without being part of the final product.

Catalysts react with one or more of the reagents to produce intermediate products, which, subsequently, lead to the final product of the reaction. Once obtained the product, the catalyst is regenerated.

For example, in:

$$A + B \rightarrow AB$$

We can assume that this reaction is very slow; however, in the presence of a compound K, the reaction is accelerated due to the formation of more-reactive intermediate compounds.

$$A + B + K \rightarrow AK + B$$
$$AK + B \rightarrow AB + K$$

As we see, at the end of the reaction, we obtain K again.

The concept of catalysis can be complex, so let us limit ourselves to know that, in the presence of certain compounds, the reactions can be faster.

Catalysts can be liquid, gaseous or solid. The only condition that should meet is that they must be part of the reaction but not of the final product.

They are very important compounds in the development of a reaction, and they make possible the production of compounds that would otherwise be impossible to achieve.

In 1933, chemists Reginald Gibson and Eric Fawcett obtained polyethylene (plastic bags) by subjecting the ethylene at a pressure of 1,400 bars and 170 ° C. These conditions are excessively harsh and cause the production of polyethylene to be unfeasible.
A few years later, Karl Ziegler and Giulio Natta were able to produce polyethylene under conditions much more compatible with the needs of the industry by using metal catalysts. This discovery earned them the Nobel Prize in 1963.

In the absence of catalysts, much of the materials that you know today would not exist.

A little bit of physics

Uff, this is a tangle! You may say. I will try to be brief by outlining the physical methods that we will use while making fuel.

The processes that you will see now do not include a change in the structure of matter, and there are no chemical changes. They are intended to separate compounds from others according to our need.

Solid Decantation

Besides of having an unattractive color, the used oil that you will use to produce biodiesel has many particles in suspension. Remember that you fried breaded filets in there and half of the bread was left in the pan.

We will use decantation to get rid of all these impurities. You will have to leave the oil resting as much time as possible (few days), and you will see that all the particles in suspension have gone to the bottom of the recipient due to gravity.

This is a process whose speed will depend in part on the viscosity of the oil, so it's a matter of patience. Gravity will take care of everything.

Decantation and separation of phases

The separation of phases occurs between two fluids of different densities, such as water and oil. If you put them together in a glass, you will see that they will separate from each other eventually, leaving the water below and the oil above in the glass. This happens because the water is denser and does not mix or react with the oil.

We will use this phenomenon to separate the products of the reaction. The products that you will obtain inside the reactor will not be combined and their density difference is large enough to allow separation in a relatively short time.

Distillation

In a distillation, you separate two liquids which are dissolved one into another. If the two liquids are not different enough, it will be impossible to separate them by decantation; then, what we'll do is to heat them, evaporate one of them and collect it in another site.

This process was used at the time of the Dry Law to manufacture liquor. In our reaction, one of the resulting products will have a high concentration of methanol. If you want to recover it to re-enter it in the following reaction, you should use a distillation process.

This is a delicate process, and it may not be necessary to perform it until another time. Right now, we simply name it and have it present for a future use.

Before making biodiesel

PLEASE, before you start combining chemicals, it is better for you to read (a couple of times) the safety data sheets of products you will use. There is a copy of them at the end of this manual, and if you cannot find some of them, you may download them at http://www.msds.com

Get yourself a gown, goggles and latex gloves. The products that you are going to work with may burn you badly. Additionally, it is necessary that you work on a ventilated site given that methanol vapors are toxic. It is highly recommended that you use a mask suitable for methanol exposure, so any store of work safety equipment will know how to advise you about this. Also, you must have a fire extinguisher properly loaded and running. FOR ANY REASON, DO NOT SAVE ON SAFETY.

Needed reagents to make biodiesel

As I mentioned before, biodiesel is a mixture of different fatty acid esters. To make biodiesel, we will need to transform the frying oil into these fatty acid esters. This transformation is accomplished through a reaction called transesterification, which we have talked somewhat.

To perform a transesterification, we will need an ester, which in our case will be the oil (oil is a triester of glycerol), an alcohol and a catalyst. We will help the reaction with agitation to increase the contact points between reagents and temperature in order to increase the speed of the reaction.

The reaction is like:

Oil + Methanol → Biodiesel + Glycerol

As catalyst, we will use sodium methoxide ($NaOCH_3$).

The oil that we will use is frying oil. Chemically, we can describe it as a triester of glycerol. Its structure would be like:

Oil – fatty acids and glycerol ester

Where -OCOR are the different fatty acid chains. They does not need to be equal to each other, and can vary according to the type of oil used.

The alcohol that we will use will be methanol. One of the reasons that we are using it is its low price because it is the cheapest alcohol you can find. In addition, the transesterification speed decreases as the size of the alcohol increases. The reaction with methanol is much faster than with ethanol and faster than with propanol.

As a catalyst, we'll use sodium methoxide that we will prepare by dissolving sodium hydroxide (NaOH) in methanol because, like the methanol, the sodium hydroxide is very cheap. Even though we could use potassium hydroxide (KOH), this is more expensive, and we would need more quantity because its molecular weight is higher.

An easy recipe - One step reaction

The production of biofuel requires practice; therefore, we will start producing biodiesel with a simple method.

We will need:

- 1.000 ml of oil (new, preferably sunflower)
- 170 ml of methanol (CH_3OH)
- 5 g of sodium hydroxide (NaOH)
- A glass container with lid, preferably clear.

Let´s start:

1) Dissolve NaOH in methanol into the glass container.

The NaOH does not dissolve easily in methanol; thus, to dissolve it, you must agitate the mixture for a while. It will take long, but it will dissolve eventually.

Be careful not to breathe methanol vapour since it is toxic. Take my advice and buy yourself a mask with filters. Also, do not touch it!

2) Heat the oil to about 50 ° C. Control the temperature with a thermometer and try not to go over that temperature.

3) Carefully, pour the oil inside the glass container that contains the methanol / sodium methoxide mixture, cover it and agitate it for 10 seconds.

4) Open the lid carefully in order to not breathe or touch the vapors that could escape. Close it again and agitate again for 10 seconds. Do the same procedure for 10 minutes.

5) Leave the container uncovered in a corner and return an hour later to see it again. You will see that two layers are formed: The upper layer is biodiesel and the bottom layer is glycerol.

6) Separate biodiesel by using a syringe, for example.

7) Let the biodiesel stand for a week in an open container. It is ready when it does not look turbid.

It's done; you have made your first transesterification! Now, let's look at what happened and how to improve this biodiesel since it is still early to use it in the automobile.

What has happened here?

Let's look at the overall reaction and then let's go step by step:

Oil + Methanol → Biodiesel + Glycerol

1) When you mix sodium hydroxide with methanol, sodium methoxide and water was formed.

$$NaOH + CH_3OH \rightarrow NaOCH_3 + H_2O$$

Even though I haven't said it to you before, you should know that there is an excess of methanol in relation to the amount of NaOH that you dissolve and therefore, that liquid that you have obtained in the first container is a mixture of sodium methoxide and methanol (our catalyst and one of the reactants, respectively).

2) Then, you heated up the oil. This is because the heat also accelerates the reaction, and you will have to wait less to reach the end of the process.
You have to be careful not to go over 55 °C because more heat will evaporate the methanol. And if the methanol evaporates, we will have no reaction.

3 & 4) By mixing sodium methoxide, methanol, and oil is when transesterification occurs. And it is when fatty acids that make part of the oil are separated from the glycerol, and they are esterified with methanol. These esters of methanol are biodiesel.

Let's look at the reaction:

| Oil (Fatty acid ester) | Methanol | Biodiesel (Fatty acid methyl ester) | Glycerol |

5 & 6) By allowing the mixture to stand and due to the different densities of biodiesel and glycerol, the two products are separated. This is a process that takes time; hence, it is important to be patient. Then, you must keep the upper part.

7) Do you remember the methoxide formation reaction? In that reaction, water was formed, and it is still inside of the bottle. It is the water of this reaction that makes biodiesel that you just separated to look turbid. By leaving the biodiesel several days in an open container, you get the water to evaporate and only the biodiesel stays in the container.
You could also warm up the biodiesel and start agitating to evaporate the water faster, but this leads to unnecessary and undesirable polymerization and degradation.

Everything that you have done is only part of the biodiesel manufacturing process, perhaps the most important part but not the only one as there are other important parts that are essential in order to avoid problems in the reaction or internal engine damage later on.

You will see next, indeed, the transesterification process modified in a way that you can get more product and therefore, lower biodiesel contamination by the reagents.

As we seen before, at the end of the reaction, there will be residual oil and methanol to be distributed between the phase of biodiesel and glycerol. A greater part of the methanol will remain in the phase of glycerol because it is more soluble there, but the oil prefers to be dissolved in biodiesel.

The presence of non-reacted oil in the biodiesel is undesirable because it is related to the presence of free fatty acids and other compounds that can produce polymers that obstruct pipes, filters and nozzles when they get into the engine.

On the other hand, pollution of biodiesel with methanol is also undesirable. The presence of methanol reduces the flashpoint of fuel and makes it more dangerous to handle.

Improving the process – Two step reaction

We talked few pages earlier about equilibrium reactions and how we had all compounds in the reactor at once at a given time.

At the end of our reaction

Oil + 3 Methanol → 3 Biodiesel + Glycerol

We will have the four components present at the same time and complying with the relationship:

$$constant = \frac{[Products]}{[Reagents]} = \frac{[Biodiesel]^3 [Glycerol]}{[Oil][Methanol]^3}$$

Where [Biodiesel], [Glycerol], [Oil], [Methanol] are the concentrations of biodiesel, glycerol, oil and methanol respectively (for example, in mol/l).

To obtain a greater amount of biodiesel, we have to add an excess of one of the reagents. We'll use an excess of methanol to be practical.

In order to maintain the relationship constant by taking into account the addition of methanol, the reaction must generate more biodiesel.

This is fine, but we want to squeeze the oil and get as much biofuel as it can. So, while the reaction is performed, we will separate the glycerol. By removing glycerol, the relationship is unbalanced, and the reaction should continue generating biodiesel again.

Let's take a look at this, in a practical way.

We will need:

- 1.000 ml of oil (New, sunflower oil is fine)
- 170 ml of methanol (MeOH)
- 5 g of sodium hydroxide (NaOH)
- A glass container with a lid, clear is preferred
- A couple of extra containers

The process:

1) Dissolve the NaOH in methanol in the glass container, and once dissolved, separate approximately 40 ml of the mixture into another container.

2) In a pan, heat the oil to around 50 °C (control the temperature with a thermometer and don't go over that temperature).

3) Pour the oil carefully into the glass container that contains the largest amount of methanol/sodium methoxide mixture, cover it and stir it for 30 minutes.

4) Leave the container uncovered and seated still for 2 hours.
5) Separate the glycerol formed with a syringe fitted with a tube that reaches the bottom of the container. Remember that glycerol is the bottom layer.

6) Heat biodiesel slightly to 50 ° C. To do this, put the container glass in "bain-marie". Start agitating and be careful not to let water to come in. It is very important not to go over the temperature and not to have moisture.

7) Pour the 40 ml of methanol / sodium methoxide that you set aside in step 1 over the biodiesel that you have.

8) Cover and stir for 30 minutes.

9) Leave the container uncovered and seated still for 12 hours.

10) Separate glycerol using a syringe in the same way you did in step 5.

11) Let the biodiesel in open container until it changes from turbid to transparent.

IMPORTANT: The biodiesel that you have done so far should NOT be used in an automobile. In fact, what we have done is to practice and see how the transesterification reaction is produced. There are other missing processes aimed at improving the condition of the fuel and make it suitable for use in your car.

Real life

We have seen how to perform the transesterification, in few words, how to transform oil into biodiesel. However, making biodiesel to use in the automobile is a longer process, and it requires that we take into account some other things.

In Figure 1, you have an outline of the entire process.

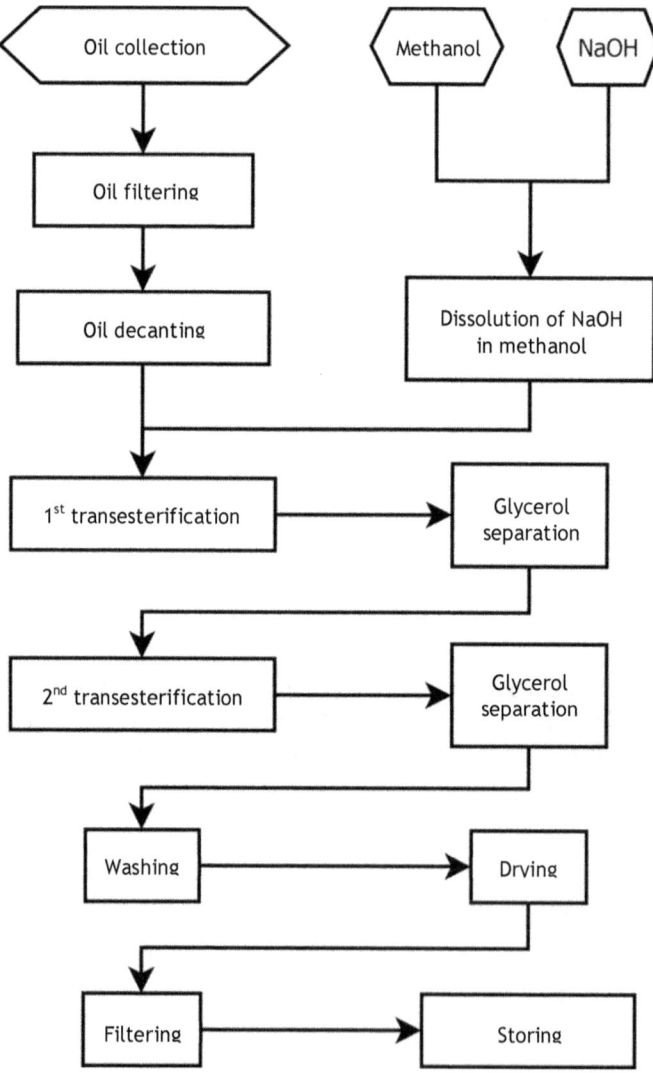

Figure 1. Required processes for the fabrication of biodiesel.

Consider now each process independently and then let's build a complete system that is able to make small batches of about 50 to 100 liters at a time.

Oil filtering

The oil that you will be utilizing is not new oil, it was used and will come to you with many impurities. Therefore, you need to filter it.

I use a wire strainer (those ones that are used to separate the pulp from the fruit juice) to transfer the oil to a storage container. With that, large particles will be separated.

I know people who use jeans fabric as a filter, which it can also work but the filtering process will be a little slower.

Oil decanting

We have removed many impurities, but we need to get rid of smaller particles. To do so, we will decant the oil for one or two weeks, so the solids will settle to the bottom of the container.

This decanting process should last at least 48 hours. Nevertheless, the more time the oil is decanting, the better will be the oil separation. It is not necessary that the tank has a conical bottom because the phase that concerns us is the upper one.

As the hours come by, due to gravity, the particles go down to the bottom of the container, and some of them also go to the surface. What matters is that the particles go either to the bottom or to the surface of the tank.

One of your worst enemies when it comes to making biofuel is water. If the oil that you are going to recycle is very used, you may have some water, and most of it will go to the bottom because the solubility of water in the oil is very low.

You should also avoid thermal flow within the tank as these will stir particles deposited at the bottom and say goodbye to all the work that we had done so far. Thus, do not mix cold oil with hot oil.

Preparation of methoxide

Let's prepare the sodium methoxide in a plastic container preferably because it is corrosive. A simple but slow way is by putting the required amount of NaOH and the corresponding methanol into a PE container. Whenever you are nearby, you can stir the preparation a little. For the next day, it is usually dissolved.

Also, you can use a tank with mechanical agitation, but be very careful that the methanol vapors do not come into contact with any spark or heat source.
You must be patient because the dissolution rate of NaOH in methanol is very low.

Beware of the reagents (NaOH and Methanol) as they are dangerous if not handled properly. Read the Material Safety Data Sheets at the end of this book.

First transesterification reation

We have filtered our oil and prepared methoxide as well, so we will mix them together with more methanol in our reactor.

The oil and methanol are not miscible with each other (do not mix). In few words, if you leave them alone, they will separate, and if this occurs, the reaction will be much slower. At this point, we must ensure that both the methanol and oil come into contact.

We will use a pump to suck up the mixture from the bottom part of the reactor, and then pour the mixture from the upper part. It is important not to let the reagents get separated.
This agitation process must last at least 45 minutes. There is no maximum time for the agitation process.

The agitation should be performed at about 50 ° C to accelerate the reaction without evaporating the methanol.

After agitating the mixture, it is necessary to transfer it to a decantation tank where we can separate the two formed phases.

Separation of glycerol

Let the mixture brought from the reactor to be seated still. After 2 hours, we will obtain two phases: the glycerol is the bottom phase while the upper phase is biodiesel.

We will separate the heavier phase (the bottom one, glycerol) in order to "move to the right" our reaction.

Second transesterification reaction

After removing the glycerol, we return to fill the reactor with biodiesel and add more methanol and catalyst to continue the reaction.
We maintain the agitation at a temperature of 50 ° C for other 45 minutes and transfer the mixture back to the decanter.

Separation of glycerol

This time, we will get less glycerol than the last time, so we will leave the mixture to decant for 24 hours minimum to ensure that we eliminate as much glycerol as possible.

The mixture will cool off, and, by being already at room temperature, no thermal flow will be inside the decanter. This will help the small drops formed to reach the bottom of the recipient faster.

By separating the heavier phase, we practically eliminate all glycerol formed that is not dissolved in the biodiesel.

We must wash with water the biodiesel to get rid of glycerol, methanol and sodium methoxide that are dissolved in it.

Washing

A common feature of the glycerol, methanol and sodium methoxide is that they are more soluble in water than in biodiesel.

If you have one or more impurities dissolved in a liquid (in our case biodiesel), it is possible that you can remove them if you get them to dissolve in another liquid that has higher affinity with them. In our case is water.

It is advisable that before washing, do it with a small amount of biodiesel that you took from the decanter. If this small sample gives you problems, so will be the biodiesel from the decanter. Emulsions that are very difficult to break in some cases can be created and can bother you during the entire process.

Once the small-scale test is done, we will add a certain amount of water in our decanter and stir for 10 minutes, so the transfer of these impurities will be faster. Let it stand for 2 hours.

The water goes down with a whitish color. Then, we remove the bottom phase from the decanter.

We repeat the operation three times. Water will come out more crystalline every time.

WARNING! THIS WATER IS NOT SUITABLE FOR CONSUMPTION; THUS, GET RID OF IT AS SOON AS POSSIBLE.

REMEMBER: It is possible that soap will be formed due to the presence of fatty acids (fatty acids + sodium + water = soap). If soap is formed, it is a sign that the oil was severely damaged or you've went over with NaOH. I have encountered with this few times, if this is your case, you have a problem.

For oils with a high acidity, which means with a large amount of free fatty acids, we will use another method that we will see later.

Drying

We already have biodiesel without methanol and without sodium methoxide, but with water. So, the biodiesel must be dried.
The simplest method is to heat the biodiesel and start agitating. By agitating, water evaporates more easily, but we can damage the biofuel. Hence, forget about this method.
Another method you see online is moving air as bubbles from the bottom of the decanter. By doing so, air bubbles will assimilate the water dissolved in the biodiesel and will take the moisture out from the liquid.

This process is quite slow as the ability of bubbles to retain humidity is not very large. Estimate that you will have to do it for at least 48 hours. Also, you favor the oxidation of biofuel, so we will not use this method either.

What we will do is to use a process called adsorption (do not be confused with absorption) by which the impurities are "stuck" to a support, usually porous.

Some home biodiesel producers use oak chips as filler in adsorption columns with a lot success. You can use those chips or commercial products that also include some additional material that helps to get a better retention of impurities.

We are talking about a column filled of the adsorbent material into which the wet biodiesel goes through. The speed of this step determines the amount of retained water. The slower is the process, the better.

Once it is dried, the biodiesel goes from turbid to transparent with a slight yellow color.

Biodiesel filtering

Finally, and although we filter the oil before starting work with it, we need to filter it again to remove the smaller particles which were not harmful to our processor but they are for the car's engine. I usually use a 5 micron filter.

Along with the 5 micron filter, it is also advisable to use a diesel filter as the one the car has.

I know people that install these filters into the car, so they filter particles that reach the biodiesel from the oil, and also filter the particles that are within the power drive circuit and, because biodiesel is more solvent than diesel, they can come off and get to the injectors and damage them.

I saw this system functioning successfully in a TDI engine. These engines are more sensitive to impurities of fuel. The only downside is that it forces the fuel pump to overcome the pressure drop of the additional filter and possibly may shorten its life.

Reactor building

As you saw, the topic becomes complicated, but we can still handle it. In the following pages, you will see how to build a reactor to produce batches of 100 liters of biodiesel. We will leave the biofuel ready to use in our automobile, blended with diesel or pure.

To build our equipment and have it ready to produce biodiesel, we will need to evaluate and define each of the tanks and reactors that we will use.

The equipment that I present here is designed to work without having to touch the reagents or reaction products. Reactor heating can be made of electric heaters or a closed circuit in which a hot liquid circulates such as heat boilers that operate in some houses.

I highly recommend professional help, especially in regard to electrical connection. ALL deposits and materials that have electrical supply must be protected with an earth wire, differential and thermal switches. We are working with flammable liquids, so we do not want anyone to be electrocuted or burn the whole equipment due to a spark.

As possible, use the material prepared to work in unstable environments. You will prevent many troubles.

The reactor is closed and has a vent valve. Do not load the reactor with that closed valve since you would be increasing the pressure inside, and this can lead to an explosion. The same applies to the unload process of the reactor since you would be creating emptiness and damaging the system.

Here I present a wiring diagram of the various elements of our biodiesel manufacturing equipment, and as you see, it is designed to work with a single pump. By opening and closing the different valves, you can direct the fluids anywhere.

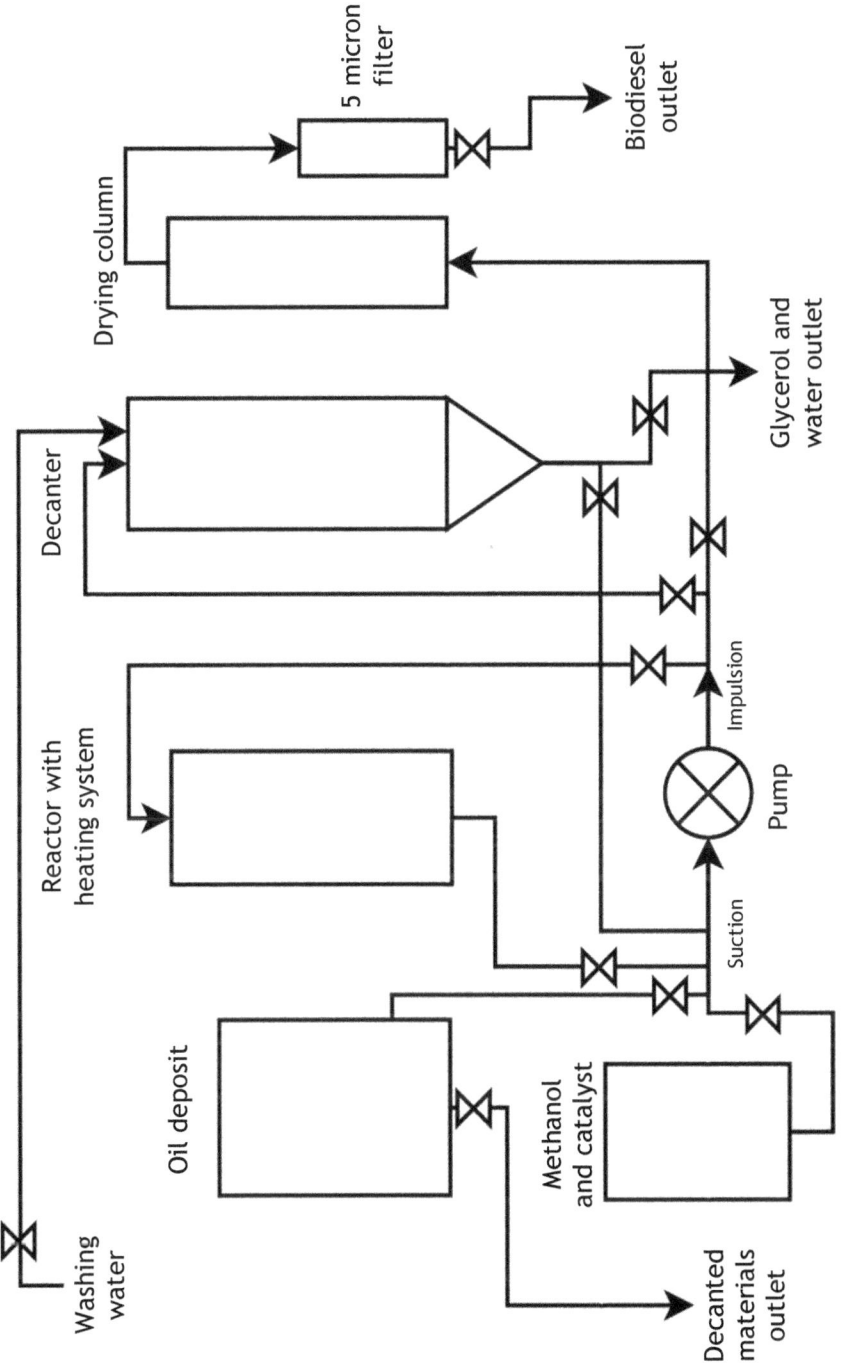

5 micron filter

Biodiesel outlet

Drying column

Decanter

Glycerol and water outlet

Reactor with heating system

Impulsion

Pump

Suction

Oil deposit

Methanol and catalyst

Washing water

Decanted materials outlet

35

Pump and connection module

We will use a pump to move the various liquids into or out of the reactor. This pump must be self-suction and made of stainless steel to avoid corrosion.
As possible, you should also be prepared to work in unstable environments, the rest of the material that follows should be connected to ground and protected with differential and thermal switches.

The pump will be connected to what we will call connection module, which are a series of keys attached to a preferably stainless steel or polypropylene pipe.

There are polypropylene (PP) pipes in the market with interior made of glass fiber. They are used for hot water and are welded together by heat. They are a very good alternative to stainless steel pipes because their price is lower. They are usually green and have longitudinal lines in other color.

In figure 2, you will see how to configure the connection module.

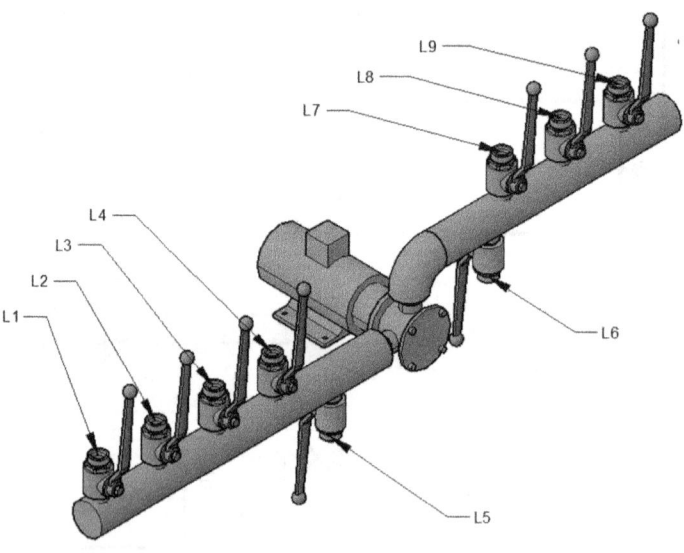

Figure 2 – Connection module.

Storage tank and oil decantation

Our storage tank will have a minimum of 200 liters of capacity; therefore, we will have oil for a couple of batches. The collection of the oil is usually the most complicated part and thus, one of the slower processes.

The material that the oil tank is manufactured with can be polypropylene, polyethylene, steel or stainless steel. We will have no corrosion problems here.

The entry of oil to the tank will be at the top, where we can place a filter to separate the larger particles. The outflow of oil will be made by a piston placed in one of the walls of the tank at a few centimeters above the bottom of the tank, so when you open the outgoing flow, it will not drag solid particles.

It is recommended to make a few marks inside the tank or, better yet, on a transparent tube, in order to know how many liters of oil we're loading. It is necessary to measure with certain accuracy the amount of reagents that we enter to the reactor.

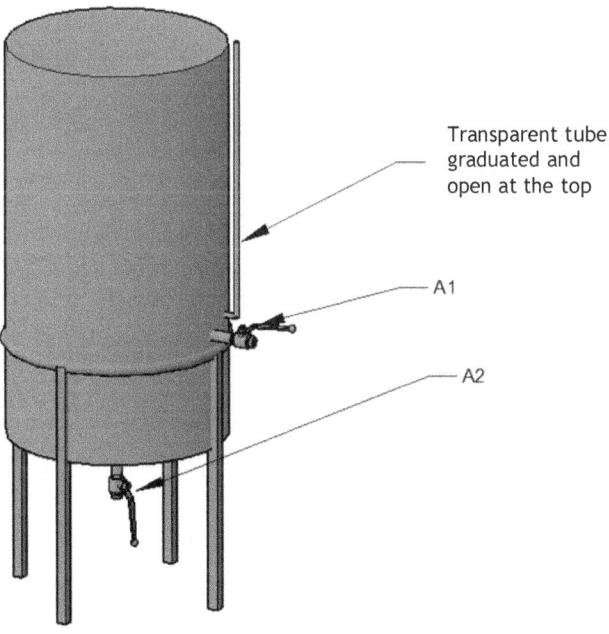

Transparent tube, graduated and open at the top

A1

A2

Figure 3 – Oil storage tank.

Methoxyde and methanol inlet

We will do the entry of the methoxide and methanol from a 20-liter-jerrycan, which is easy to handle and will likely to be the container that will be sold to you.
Make sure there are no heat sources nearby, DO NOT SMOKE or do sparks near the methanol as it is highly flammable.
Add the necessary grams of NaOH to the 20-liter-jerrycan, stir it to dissolve the catalyst, and it will be ready to be used. Very important, sodium hydroxide (NaOH) must be dissolved. Do not start until this happens.

To the L4 key of the connection module, we will adapt flexible silicon hose which its free end will be immersed into the jerrycan.

Reactor with heating

The reactor will have a capacity of about 180 to 200 liters. It will be made of stainless steel and is not necessary to have a conical base. If it has a conical base, it would not be a problem either.
Inside will occur the transesterification reaction, and for that, we must heat the mixture through an electrical resistance or, better yet, the movement of any other hot liquid.

In figure 4, you have a diagram of the reactor with electrical heating (the resistance cannot be seen), although it is also possible to install the same reactor with heating by recirculation.

Pay special attention to the valve called R4 as it will serve you to close the reactor hermetically. It prevents the evaporation of methanol and its consequent outflow to the atmosphere. In both the loading and unloading processes of the reactor, the piston must remain open to prevent explosion or implosion.
To engage an electrical resistance to the reactor, the best thing to do is screw it since it avoids glues and other rare inventions. Threaded with teflon is the safest way. You must install the heat source, either a resistor or a heater circuit, as close as possible to the bottom of the reactor.

There are some models of reactor in the market that perform the heating of the reacting mixture while passing through the pipe that goes up to the top of the reactor. In our case, the pipe called R3 will be heated. Heating the liquid flow is not a bad idea, as some home heaters work the same way.

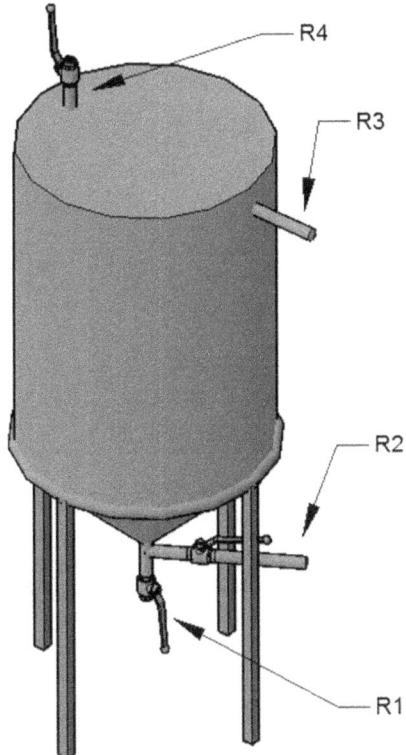

Figure 4 – Reactor with electrical heating.

You must also adapt it a thermometer to check the temperature inside. The heat source must be connected to a thermostat so they do not to exceed the work temperature. It would also be useful to have a pressure gauge to avoid overpressure problems.

Glycerol decanter

The glycerol decanter must be only that: a decanter. Therefore, it must have a conical base. It must be made of molded stainless steel, polypropylene or polyethylene, and it should be open on the top because we will install the washing shower later on.

The connection with the main module can be seen in Figure 5.

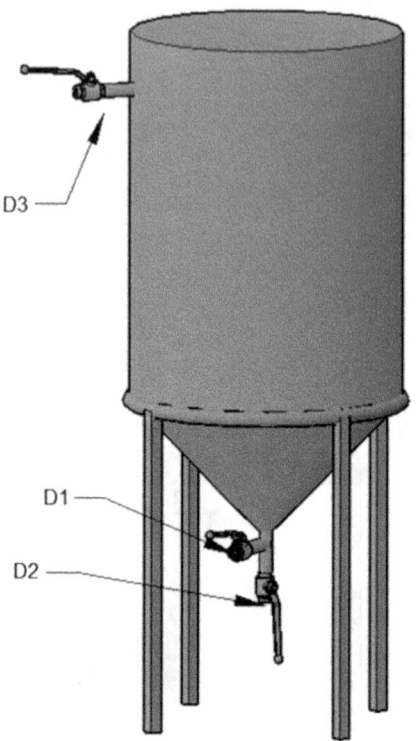

Figure 5 – Decantation tank.

Washing shower

The washing shower does not have a lot of mystery since it is very similar to those ones in the house, it need a flow meter in order to show how many liters of water we have added. This shower is connected to a water tap and must be installed on top of the decanter.

The need of the flow meter and the shower is due to the fact that it is not possible to obtain a homogeneous rain from jerry cans or containers of measurement. Flow meters are instruments fairly well known in the plumbing sector and shouldn't be very complicated to get them.

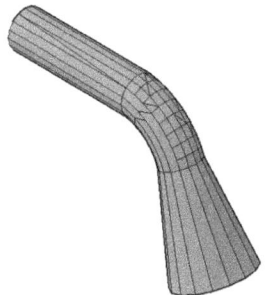

Figure 6 – Washing shower.

Adsorption column

The adsorption column is a tube made of a plastic (polypropylene or polyethylene) or metallic material and of approximately 1.5 m long and 16 cm in diameter, threaded at its ends and with respective covers. In order to avoid the adsorbent material to come out with the biodiesel, these caps have a filter that does not allow internal particles of the column to be dragged outwards.

The tube should be placed upright, and the fuel must pass through from bottom up at a slow speed. The material to fill the column is commercially available under the name "Dry wash purification media", although some house manufacturers use oak chips with considerable success. The performance of these materials is quite high and hardly affects the final cost of the product.

The column filling can be used several times, according to the manufacturer's instructions. Once completed its service life, it will be contaminated with methanol, water and glycerol.

There are people who use these chips to make them burn in a fireplace. If you do this, be sure to use them as fuel once the fire is hot because if the glycerol is not burned at a high temperature, it may generate carcinogenic and toxic gases.

In the figure below, you will see a standard adsorption column. If necessary, you can adapt other columns in parallel.

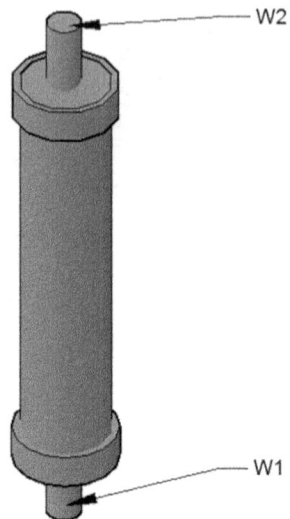

Figure 7. Adsorption column.

Microparticle filter

Use a filter identical to the one you have installed in the automobile. The idea is that if the fuel that we have made has any particles that might damage the engine, they will stop at this filter. If it passes through the filter, it will not likely damage the engine.

Install the filter on a suitable support and, with the help of the pump, pass the biodiesel through it. We will call to the inlet connection entry F1 and the outlet connection F2.

Biodiesel storage

The best for storing the biodiesel are the 20 or 25 liter jerry cans as they will allow you to manipulate them with relative ease and in the case of the biofuel is damaged for some reason, this will not affect the remaining.

The jerry cans can be preferably plastic (polyethylene or polypropylene) or metallic by default. The same jerry cans used for storing gasoline can be used for storing biodiesel.

The jerry cans should not have water or dirt because it is possible that this dirt end up in the engine and damage it. Think also that it can be the focus of bacterial contamination, and we do not want bugs inside the engine.

Module connections

In the table below are shown the modules jacks.

Jack	Connect with	Use
L1	R2	Reactor recirculation (suction)
L2	A1	Oil inlet
L3	D1	Inlet from decanter
L4		Methanol/Methoxide inlet
L5		Flushing valve
L6		Flushing valve
L7	W1	Outlet to adsorption column
L8	D3	Outlet to decanter
L9	R3	Reactor recirculation (impulsion)
A1	L2	Oil outlet
A2		Flushing valve
R1		Flushing valve
R2	L1	Reactor recirculation (suction)
R3	L9	Reactor recirculation (impulsion)
R4		Decompression valve
D1	L3	Outlet to connection module
D2		Flushing valve
D3	L8	Inlet from connection module
W1	L7	Dryer inlet
W2	F1	Outlet to microparticle filter
F1	W2	Inlet from adsorption column
F2		Biodiesel outlet

Instructions for using the reactor

Now that you have assembled the reactor following the schemes of the previous chapter, we have to make it work.

BEFORE YOU BEGIN, CLOSE ALL KEYS OF THE EQUIPMENT

Materials needed to make 100 liters of biodiesel:

- 100 liters of vegetable oil.
- 20 liters of pure methanol.
- 500 grams of sodium hydroxide >98%.
- Our reactor.
- Some containers.

Oil filtering and decanting

Pass the oil through a sieve or mesh to eliminate the largest impurities, and then, leave it for a minimum of 48 to 72 hours in the tank to decant the solid particles.

The decanted oil now must enter into the reactor. To do so, open the keys A1, L2, L9, and R4 and turn on the pump connection module (M1). You can see how many liters enter into the reactor through the graduated tube. Stop the pump when 100 liters have entered into the reactor.

Close all keys again. Later on, you will notice that it is not necessary to close them all, but we will make so to avoid errors.

Catalyst fabrication

24 hours before you begin, add 500 gr. of NaOH to a jerry can with 20 liters of methanol and let it dissolve into the methanol. Stir occasionally the mixture to accelerate the process. DO NOT START IF THE NaOH IS NOT DISSOLVED COMPLETELY.

Once the NaOH is dissolved, you have to separate 5 liters in another container with the purpose of using it in the second stage of the transesterification.

In the position L4, you have a hose that will help you to empty the jerry can.

For that, you have to open the keys L1, R2, L9 and R4. Then, start the engine. At this time, you will be re-circulating the content of the reactor. Open slightly L4 to lower the level of the mixture of methanol/methoxide. Once you have emptied the 15 liters, close L4.

First transesterification reaction

You already have the oil and methanol re-circulating. Now, turn on the heating system of the reactor (either a resistor or other system). Control with the thermometer that the temperature does not go above 55 °C. You can close R4 to prevent methanol vapors that may occur from escaping the reaction, but THIS COULD CREATE A REACTOR OVERPRESSURE. Thus, use a pressure gauge if you are going to do this and do not overload the reactor beyond the limit established by the manufacturer.

Once the temperature reaches 55 °C, turn off the heater and allow the content to react by re-circulating for 45 minutes approximately. After this, turn off the engine and close all keys.

First decanting

It is time to decant the glycerol in order to make progress in the reaction. To do so, open up R2, R4, L1, L8, and D3 and turn on the pump M1. When the reactor is empty, shut off the pump and close all keys.

Leave it to decant for 2 hours and discard the lower layer (glycerol) through D2. When you see biodiesel coming out, close D2.

Once separated the glycerol, you must return the biodiesel back to reactor by opening D1, L3, L9, and R4 and turning on the pump M1. As always, the rest of keys must be closed.

Second transesterification reaction

Once the decanted biodiesel has entered the reactor, the engine is turned off and all keys are closed. Then, this reaction proceeds in the same manner as in the first reaction. That is:

Open the keys L1, R2, L9 and R4, and then start the engine. At this point, you will be re-circulating the content of the reactor. Open slightly L4 to suck up the last 5 liters of methanol / methoxide mixture. Once entered the 5 liters remaining, close L4.

Turn on the heat and let it circulate for 45 minutes.

After the time required has elapsed for the second reaction, shut off the pump, close all keys and pass the mixture back into the decanter again.

Second decantation

Open R2, R4, L1, L8, and D3 and turn on the pump M1. When the reactor is empty, shut off the pump and close all keys.

Leave it to decant for 2 hours and discard the lower layer (glycerol) through D2. Once removed the glycerol, you must wash the biodiesel.

Washing

You will wash with water. Up to this point, the presence of water was fatal; however, if everything has gone well so far, you will not have any problem.

At this point, you will realize whether your reaction went well or not. To prevent you from destroying everything that you have done up to now, it is recommended that you do this step on a small scale first. Do not wash the biodiesel if it does not pass the test 27/3 that I will talk about on a few pages later.

Using the shower with a flow meter, let drop a flow of 30 liters of water (as thin as possible) on the fuel. As the water will go to the bottom, you will have to stir with the pump M1. For this, with all keys closed, open D1, D3, L8 and L3. Then, turn on M1.

Stir for about 10 minutes, turn off M1, close the keys mentioned above and set aside the mixture to rest. Discard the rinse water that will be turbid and white through D2.

Repeat this procedure three times or more until you see the water comes out clear. It is very important that the water is transparent and with pH of about 7 (you will understand what pH is, later on). There are pH-meters available in the market for a very low price that you can use to make this measurement.

Drying

With the washing, you have gotten rid of many impurities, but you still have a very important impurity: water.

Dry out the biodiesel by passing it through the adsorption column. To do so, with all keys closed, open D1, L3 and L7 and then, turn on M1.

If the pump M1 has too much power, the biodiesel may come out from the column too fast and it will not dry out. If so, you may need to pass fuel directly from the decanter using only the force of gravity.

You can also make the column longer, or put two of them in series, so the contact time of the liquid with the column filling would be greater.

Then, biodiesel will pass through the filter of micro-particles and will be ready to use. Check the filter periodically searching for water or bacteria. You may damage all work at the last minute.

The biodiesel that comes out of here must be crystalline, colored but without any turbidity.

Bottling

The biodiesel as any organic product is sensitive against bacteria attack. Fighting against them is not among the priorities of this book, so what I recommend is to put your fuel in 25 liter jerry cans firmly closed and consume it within 6 months. Do not leave it exposed to extreme temperatures or sunlight.

Needless to say, jerry cans must be cleaned and dried. If they were filled with biodiesel before, it is OK to reuse them; nonetheless, avoid water at all cost.

Other things you should know

Titration

There is something called pH, which is the measure of acidity and technically is equal to

$$pH = -\log([H^+])$$

Where $[H^+]$ is the concentration in mole per liter of hydrogen ions. All the values less than 7 are acidic, and the pH values greater than 7 are basic or alkaline. The pH value of 7 corresponds to the pure water and is considered neutral.

When the cooking oil is heated in the presence of water (the water from foods), a hydrolysis process that generates free fatty acids occurs. This is the reverse reaction of the esterification.

| Oil | Water | Heat | Diester | Free fatty acid |

(Fatty acid ester)

The free fatty acid dissociates into:

$$\text{R-CCOH} \rightarrow \text{R-COO}^- + H^+$$

The pH value will give a measure of the amount of free fatty acids in the oil.

These free fatty acids form soaps when reacted with sodium catalyst and the water that is formed when dissolving NaOH in methanol. Then, we will have two problems: Soap in the mixture which is difficult to separate and less amount of catalyst, which will slow the reaction.

How to calculate the amount of catalyst

The amount of free fatty acids present will depend on the oil (which may contain them already before be used), and the use that has been given to that oil; therefore, we cannot predict the amount of catalyst needed for the reaction.

It is necessary to titrate the oil, which is measuring the acidity of it. Once we know how much free fatty acid exists, we will be able to compensate the loss of catalyst.

The titration procedure is as follows, and we will need:

- 10 ml of oil.
- Phenolphtalein 2%.
- A 0,1% solution of NaOH in water.

The first thing we will do is to prepare a 0.1% solution of NaOH. We must weigh 10 grams of dry NaOH (the most pure that we may find) and dissolve it into 1 liter of distilled water.

Once dissolved, we take 100 ml of this mixture and dilute it in a new container with 900 ml of distilled water. We have 1000 ml of standard solution. We will call it solution A.

Now, we prepare the oil. To do so, we measure 10 ml of oil and mix it in a clear glass beaker with 100 ml of isopropyl alcohol and two drops of phenolphthalein. Stir well to mix.

Later, add solution A with a burette, or simply with a graduated syringe, drop by drop. Stir constantly and continue to add drops of solution A until the oil mixture changes color (to pink). Here is when you should stop. Keep stirring until it changes color again and re-add solution A.

You should do this until the mixture do not longer go back to the original color. Write down how many milliliters of solution A you spent when it happens.

To calculate the amount of NaOH that you must add to the reaction, use the following formula:

$$C - \left(\frac{ml\ titration}{10} + 5 \right) * [Oil\ litres]$$

C is the amount in grams of NaOH that you must dissolve in methanol.

*In the reactions that we saw throughout the book, we believed that the amount of free fatty acids was 0; therefore, the number of grams of NaOH that we had added to 100 liters of oil was 500 g. (5 G/L * 100 l)*

Acid esterification

When the amount of free fatty acids is very high, the danger of forming emulsions (soaps) is also high. The "trick" of adding more catalyst does not work in these cases because we are also adding more water and thus favoring emulsion.

What you need to do is convert these fatty acids into biodiesel with an esterification reaction in an acidic medium.

DO NOT USE THIS PROCEDURE UNLESS STRICTLY NECESSARY AND IF YOU HAVE DOMINATED ALL OF THE ABOVE. IF YOU STILL WANT TO DO IT, BEGIN WITH SMALL QUANTITIES.

To esterificate free fatty acids of the oil, we will need:

- 1 liter of oil.
- 1 ml of sulfuric acid 98%.
- 100 ml of methanol.

Heat the oil to 35 °C, Once this is done, mix the oil and methanol with constant stirring. Then, while stirring, add the sulfuric acid very slowly. Keep stirring for 2 hours.

BE CAREFUL WITH SPLASHES BECAUSE THE MIXTURE CAN BE REALLY HOT

Take a sample and make a titration. The result should be less than or equal to 4 ml of 0.1% NaOH solution. If this is not the case, keep stirring until it reaches that value.

Using the measure of the last titration, correct the amount of catalyst and begin the transesterification in two stages as always.
Finally, you will get more biodiesel than before and also more water. Unless that your oil is very bad, you will not need to use this procedure, but it is good to know that it exists.

Quality controls

Quality controls are necessary to judge whether the process has gone fine and to make sure you don't broke your car in the middle of the road.

There is a standard in Europe to value whether a biodiesel is suitable to be used safely in automobiles, which is the UNE 14214. In the U.S., is the D6751-02.

Unfortunately the tests specified there require of equipment that you will not have at home, so we must ensure our quality in a more basic way.

In any case, if any of these tests is negative, it is not advisable to use that fuel.

Water in oil

With this procedure, you can find out if your oil has water and how much quantity.

You will need:

- A digital scale (accurate as possible).
- A thermometer capable of measuring up to 180 ° C.
- A metal container.
- A source of heat.

Weigh 1,000 g of oil in a metal container. Heat it to 120 ° C for one hour, and stir it if possible. After that time, carefully reweigh it.

The difference in weight will give the amount of water present by the formula:

$$\%_{of\ water} = \frac{W_{initial} - W_{final}}{10}$$

You should not go over 1-2%

Washing Test

This test will tell you if you have soap in the biodiesel. If so, it is very likely that you have problems in the engine later on. The presence of soap in biodiesel is indication of presence of water, an incomplete reaction and possibly a false measurement of the catalyst.

You will need:

- A container with a lid.
- 150 ml of distilled water.
- 150 ml of unwashed biodiesel (without glycerol).

Fill a container with biodiesel and then add distilled water. Cover it and stir it strongly for 10 seconds. Let it stand.

After 30 minutes, you should see two distinct layers. If not, your oil probably had water, and/or many free fatty acids, or the reaction was incomplete so DO NOT WASH THE REMAINING BIODIESEL.

27/3 Test

This test will tell you if the transesterification reaction was performed completely or, if the oil still needs to react. This test must be done before washing.

You will need:

- A container with a lid.
- 27 ml of methanol.
- 3 ml of washed and dried biodiesel.

Fill a container with 3 ml of biodiesel and add 27 ml of methanol. Cover it and stir it strongly for 10 seconds.
If biodiesel is completely dissolved in methanol, it means, that the transesterification reaction is complete. If on the other hand, drops remain on the bottom, the reaction has been incomplete.

The drops at the bottom are made of oil (insoluble in methanol). If you have oil, the reaction is not complete. Do not use this fuel. Take it to the reactor and re-do the transesterification.

Obtaining oil

Your sources of oil will be your greatest asset. Currently, it is not easy to get used cooking oil because there are many companies dedicated to pick it up in bars and restaurants to sell it to large biodiesel plants. But there is still hope. Remember that your oil needs are not as large, and you may get it with a little bit of cleverness.

Some ideas:

- Family and acquaintances: Your neighbors and family members are probably keeping their used oil to take it to the appropriate container. Avoid them this problem by picking it up for them. When they find out that you want the used oil for an ecological cause, they will help you without any problem.

- Bars and restaurants: Although many of them are reluctant to "give you" their used oil, there is always one that does. Do not get discouraged and continue entering at each bar or restaurant you find.

- The land field: The oil comes from the land field. You may be able to get into an agreement with the owner of any sunflower, olive or other crop field, in exchange for biodiesel. He may give you a few liters of oil. The oil producers, due to quality reasons, must reject many liters of oil that may serve you.

- Algae: Currently, there are researches about obtaining triglycerides from algae. The idea is somewhat simple: You cultivate certain species of algae, and then you press them to obtain the triglycerides inside to use in place of oil in the standard process of manufacturing biodiesel. If you have time and desires to investigate, this is the future because you can get very high yields.

- Do it yourself: If you have an abandoned lot, you can grow your own raw material, press it and get this precious oil. According to the weather in which you are located at, you are able to plant sunflower and soya, or if your environment is very hostile, you can think about using Jatropha Curcas, a plant that actually grows in any environment and is increasingly being used to produce raw material. The theoretical yield of Jatropha is about 1800 liters of oil per hectare.

Using biodiesel

We have prepared biodiesel, and it is time to feed your car with it. Keep in mind that not all automobiles are ready to run with a fuel other than diesel.

Some points you should control:

- Biodiesel can contaminate engine oil. Some automobile manufacturers recommend changing the oil more frequently to avoid this drawback.
- Biodiesel cannot be used in engines that perform a post-injection, designed to clean the filter from particles.
- It is possible that some tubes begin to "sweat" or directly break. You must change them for tubes made of biodiesel resistant materials.
- Inside of the tank, many automobiles carry an additional pump to send fuel to the engine. Check if it works correctly and if their parts are not damaged.
- You must change the diesel filter during the first miles with biodiesel. The dirt deposits left by fossil diesel peel off and end up in the filter.
- Check that the fuel pump has no leaks.
- Biodiesel gets very thick when cold, so do not use biodiesel at 100% below 20 ° C. If you use it at lower temperatures, it is better that you mix it with diesel fuel if you do not want to damage the engine. Do not pass from the 50%.

You have to know that the transition from diesel to biodiesel should be gradual to avoid problems.

I often recommend to new users to start mixing the biodiesel with regular diesel in increasing proportions. For example, 10% of biodiesel and 90% of diesel for the first deposit, 20% biodiesel for the second and so on until reach 100%.

In www.biodieselcasero.com you can find some additional articles that can help you. Read them before you start.

Safety sheets

Methanol

FORMULA CH$_3$OH
MOLECULAR WEIGHT 32.4 g/mol
SYNONYMS Ethyl Alcohol, Spirit, Alcohol

HAZARD IDENTIFICATION

Routes of Exposure: Methanol may affect the body either through ingestion, inhalation, or contact with the eyes and/or skin.

Summary of Acute Health Hazards Ingestion: Toxic. Symptoms parallel inhalation. Can intoxicate and cause blindness. Poison – may be fatal if swallowed. Even small amounts (30-250 ml methanol) may be fatal. May cause central nervous system disorder (e.g., narcosis involving a loss of coordination, weakness, fatigue, mental confusion and blurred vision) and/or damage.

Inhalation: A slight irritant to the mucous membranes. Toxic effects exerted upon nervous system, particularly the optic nerve. Once absorbed into the body, it is very slowly eliminated. Symptoms of overexposure may include headache, drowsiness, nausea, vomiting, blurred vision, blindness, coma, and death. A person may get better but then worse again up to 30 hours later.

Skin: Prolonged contact with the skin may cause reddening and defatting of the skin and may aggravate an existing dermatitis.

Eyes: May cause mild redness and swelling of the conjunctiva, with transient superficial injury of the cornea.

Summary of Chronic Health Hazards: N/A

Signs and Symptoms of Exposure: Methanol may affect the body either through ingestion, inhalation, or contact with the eyes and/or skin. Effects of Overexposure: Long-term repeated exposure to high vapor concentrations (greater than 1000 ppm may produce impairment of vision.

Medical Conditions Generally Aggravated by Exposure: Persons with pre-existing skin disorders, impaired liver function, impaired renal function, or pre-existing eye diseases might have increased health risks working with methanol.

Note to Physicians: Treatment should include the following: hemodialysis; the intravenous administration of ethanol (10 ml per hour) to interfere with the metabolism of methanol; and the administration of sodium bicarbonate to correct acidosis. Keep under medical supervision for at least 48 hours.

FIRST AID MEASURES

Ingestion: Give two glasses of water and induce vomiting. GET MEDICAL ATTENTION IMMEDIATELY. Do not make an unconscious person vomit. Inhalation: Move the exposed person to fresh air at once. If breathing is difficult, administer oxygen; if breathing has stopped, perform artificial respiration. GET MEDICAL ATTENTION IMMEDIATELY.

Skin: Promptly flush the contaminated skin with water. If skin irritation persists, GET MEDICAL ATTENTION. Wash contaminated clothing before reuse. Destroy or thoroughly clean contaminated shoes.

Eyes: Wash eyes immediately with large amounts of water for 15 minutes, lifting the lower and upper lids occasionally. GET MEDICAL ATTENTION IMMEDIATELY. Contact lenses should not be worn when working with this chemical.

FIRE FIGHTING MEASURES

Flash Point: 53.6°F, 12°C, T.C.C.
Autoignition Temperature: 725°F
Lower Explosive Limit: 7.3% by volume in air
Upper Explosive Limit: 36% by volume in air
Unusual Fire and Explosion Hazards: Vapors formed from this liquid, are heavier than air, and may be moved by air currents. Flashback of flame along the vapor trail to the handling site may occur. Can be ignited easily and burns vigorously.
Extinguishing Media: Dry chemical, alcohol foam, or carbon dioxide.
Special Firefighting Procedures: Use water spray to cool fire-exposed containers and structures. Approach methanol fire with caution; methanol burns with an almost invisible flame in daylight. Use self-contained breathing apparatus and protective clothing.

ACCIDENTAL RELEASE MEASURES

Eliminate all ignition sources. Handling equipment must be grounded to prevent sparking. For large spills, evacuate the hazard area of unprotected personnel. Wear appropriate respirator and protective clothing. Shut off source of leak only if safe to do so. Dike and contain. If vapor cloud forms, water fog may be used to suppress; contain run-off. Remove with vacuum trucks or pump to storage/salvage vessels. Soak up residue with an absorbent such as clay, sand or other suitable material, and place in non-leaking containers for proper disposal. Flush area with water to remove trace residue; dispose of flush solutions as above. For small spills, take up with an absorbent material and place in non-leaking containers; seal tightly for proper disposal.

HANDLING AND STORAGE

Protect against physical damage. Store in a cool, dry well-ventilated location, away from any area where the fire hazard may be acute. Outside or detached storage is preferred. Separate from incompatibles. Containers should be bonded and grounded for transfers to avoid static sparks. Storage and use areas should be 'No Smoking' areas. Use non-sparking type tools and equipment, including explosion proof ventilation. Containers of this material may be hazardous when empty since they retain product residues (vapors, liquid); observe all warnings and precautions listed for the product. Do Not attempt to clean empty containers since residue is difficult to remove. Do not pressurize, cut, weld, braze, solder, drill, grind or expose such containers to heat, sparks, flame, static electricity or other sources of ignition: they may explode and cause injury or death.

Other Precautions: The reaction of methanol with nitric acid is considered hazardous not only because it is exothermic, but also because it produces methyl nitrate. Methyl nitrate reportedly can explode violently if shocked mechanically or heated. **Disposal:** At low concentrations in water, methanol is readily biodegradable in biological wastewater treatment plant.

EXPOSURE CONTROLS / PERSONAL PROTECTION

Respiratory Protection: Use MSHA/NIOSH approved self-contained breathing apparatus in high vapor concentrations.

Respirator Selection
2000 ppm: SA/SCBA
10,000 ppm: SAF/SCBAF
25,000 ppm: SAF: PD,PP,CF
Escape: SCBA

Ventilation: This product should be confined within closed equipment, in which case general (mechanical) room ventilation should be suitable. Special, local ventilation is needed at points where vapors are expected to be vented to the workplace air.

Protective Clothing: Avoid prolonged or repeated contact with the skin. Wear chemical-resistant clothing, including boots, gloves, lab coat apron or coveralls, as appropriate, to prevent skin contact.

Eye Protection: Avoid contact with the eyes. Wear chemical goggles if there is the likelihood of contact with the eyes. Maintain eye wash fountain and quick-drench facilities in work area.

Other Protective Clothing or Equipment: An eye bath, safety shower, chemical apron and boots should be available.

Work/Hygienic Practices: All employees who handle methanol should wash their hands before eating, smoking, or using the toilet facilities. Do NOT place food, coffee or other drinks in the area where dusting or splashing of solutions is possible.

STABILITY AND REACTIVITY

Stability: Stable
Hazardous Polymerization: Will Not Occur
Conditions to Avoid: Avoid heat, sparks and flame - all ignition sources.
Materials to Avoid: Alkali metals, concentrated nitric and sulfuric acids, aldehydes, acyl chlorides, strong bases, and strong oxidizers. The reaction of methanol with nitric acid is considered hazardous not only because it is exothermic, but also because it produces methyl nitrate.
Methyl nitrate reportedly can explode violently if shocked mechanically or heated.
Hazardous Decomposition Products: Burning can produce carbon monoxide and/or carbon dioxide, and formaldehyde. Irritants. Toxic gas.

TOXICOLOGICAL INFORMATION

General: Prolonged and repeated exposure to high vapor concentrations, skin absorption or ingestion of methanol may result in visual disturbances, metabolic acidosis, headache, giddiness, nausea, insomnia, gastric disturbance, dizziness, and slow breathing. There have been severe cases reported of blindness, coma and death due to the ingestion of methanol. Acute toxicity data, if available, are listed below.

Toxicity Data

Oral LD50: 5628 mg/kg (rat),
Inhalation LC-50: (rat) 87.5 mg/l 6.00 Hours
Dermal LD-50: (rabbit) 15.8 g/kg
Skin irritation: (guinea pig) moderate
Eye Irritation: (rabbit) slight
IDLH Value: 25,000 ppm,
Aquatic: 250 ppm/11 hr/goldfish/died/fresh water,
Biological Oxygen Demand (BOD): 0.6 to 1.12 lb/lb in 5 days

ECOLOGICAL INFORMATION

Oxygen Demand Data:
BOD-5: 0.76 – 1.12 g/g
BOD-20: 1.26 g/g
COD: 1.05 – 1.5 g/g

Acute Aquatic Efects Data:
96 h LC-50 (fathead minnow): > 100 mg/l
96 h LC-50 (sideswimmer): > 100 microliter(s)/1 NOEC: 100 microliter(s)/1
24 h EC-50 (daphnid): 10000 mg/I
96 h LC-50 (daphnid): > 1000 microliter(s)/1 NOEC: 100 microliter(s)/1
96 h LC-50 (ramshorn snail): > 100 microliter(s)A NOEC: 100 microliter(s)/l

Expected to be readily biodegradable.

Partition Coefficient (n-octanol/water): -0.77

DISPOSAL CONSIDERATIONS

This product when spilled or disposed is a hazardous waste (RCRA-40 CFR 261). Preferred method is incineration or biological treatment in a federal approved facility. Consult Federal or Local Authorities for proper disposal procedures.

Sodium hydroxide

FORMULA NaOH
MOLECULAR WEIGHT 40.01 g/mole
SYNONYMS Lye, sodium hydrate, white caustic, caustic soda, soda lye, soda ash, ascarite

HAZARD IDENTIFICATION

Routes of Exposure: Sodium hydroxide can affect the body if it is inhaled or if it comes in contact with the eyes or skin. It can also affect the body if it is swallowed. Summary of Acute Health Hazards

Ingestion: Corrosive! Swallowing sodium hydroxide may cause severe burns of the mouth, throat, esophagus, and stomach. Death may result. Severe scarring of the throat may occur on recovery after swallowing sodium hydroxide. Symptoms may include sneezing, bleeding, vomiting, diarrhea, fall in blood pressure. Damage may appear days after exposure. An increased number of esophageal cancer cases have been reported to occur in individuals who have scarring of the esophagus from swallowing sodium hydroxide.

Inhalation: Severe Irritant. Effects from inhalation of the dusts, mists, or spray will vary from mild irritation to destructive burns depending on the severity of exposure. Symptoms may include sneezing, sore throat or runny nose. Severe pneumonitis may occur.

Skin: Corrosive! Contact of the skin may cause skin irritation and, with greater exposure, severe burns with scarring.

Eyes: Corrosive! Sodium hydroxide is destructive to eye tissues on contact, will cause severe burns that result in damage to the eyes and even blindness. Contact lenses should not be worn when working with this chemical.

Summary of Chronic Health Hazards: The chronic local effect may consist of multiple areas of superficial destruction of the skin or of primary irritant dermatitis. Similarly, inhalation of dust, spray, or mist may result in varying degrees of irritation or damage to the respiratory tract tissues and an increased susceptibility to respiratory illness. Effects may be delayed.

Signs and Symptoms of Exposure: A physician should be contacted if anyone develops any signs or symptoms and suspects that they are caused by exposure to sodium hydroxide.

Effects of Overexposure: Sodium hydroxide is a strong alkali and is corrosive to any tissue with which it comes in contact.

Medical Conditions Generally Aggravated by Exposure: Sodium hydroxide is a respiratory irritant. Persons with pre-existing skin disorders or eye problems or impaired pulmonary function may be at increased risk from exposure, and should have limited exposure to this material.

Note to Physicians: Perform endoscopy in all cases of suspected sodium hydroxide ingestion. In cases of severe esophageal corrosion, the uses of therapeutic doses of steroids should be considered. General supportive measures with continual monitoring of gas exchange, acid-base balance, electrolytes, and fluid intake are also required.

FIRST AID MEASURES

Ingestion: Do Not Induce Vomiting. If the person is conscious, give him large quantities of water immediately to dilute the sodium hydroxide. Do not attempt to make the exposed person vomit. DO NOT INDUCE VOMITING! GET MEDICAL ATTENTION IMMEDIATELY.

Inhalation: Move the exposed person to fresh air at once. If breathing has stopped, perform artificial respiration. If breathing is difficult, give oxygen. Keep the affected person warm and at rest. GET MEDICAL ATTENTION IMMEDIATELY.

Skin: Immediately flush contaminated skin with water. If large areas of the body are contaminated or if clothing is penetrated, immediately use safety shower, removing clothing while under the shower. Flush exposed areas with large amounts of water for at least 15 minutes. GET MEDICAL ATTENTION IMMEDIATELY. Wash clothing before reuse.

Eyes: Immediately flush eyes with a directed stream of water for at least 15 minutes. Forcibly hold eyelids apart to ensure complete irrigation of all eye and lid tissue. Washing eyes within 1 minute is essential to achieve maximum effectiveness. GET MEDICAL ATTENTION IMMEDIATELY. Contact lenses should not be worn when working with this chemical.

FIRE FIGHTING MEASURES

Flash Point: Not combustible
Autoignition Temperature: Not combustible
Lower Explosive Limit: N/A
Upper Explosive Limit: N/A
Unusual Fire and Explosion Hazards: Not combustible but solid form in contact with moisture or water may generate sufficient heat to ignite combustible materials. Contact with some metals can generate hydrogen gas. During a fire, irritating and highly toxic gases may be generated by thermal decomposition or combustion. Vapors may be heavier than air.

Extinguishing Media: Foam, carbon dioxide, or dry chemicals may be used where this product is stored. Adding water to caustic solution generates large amounts of heat. Do NOT get water inside containers.

Special Firefighting Procedures: This product is not combustible. Full protective clothing and self-contained breathing apparatus should be worn in areas where product is stored.

ACCIDENTAL RELEASE MEASURES

Leaks should be stopped. Spills should be contained and cleaned up immediately. Spills should be removed by using a vacuum truck. Neutralize remaining traces of material with any dilute inorganic acid such as hydrochloric, sulfuric, nitric, phosphoric, or acetic acid. The spill area should then be flushed with water, followed by liberal covering of sodium bicarbonate. All clean-up material should be removed and placed in approved containers, labeled and stored in a safe place to await proper treatment or disposal. Spills on areas other than pavement (dirt or sand) may be handled by removing the affected soils and placing in approved containers. Avoid runoff into storm sewers and ditches which lead to waterways. Persons not wearing protective equipment and clothing should be restricted from areas of spills until cleanup has been completed.

HANDLING AND STORAGE

Prevent possible eye and skin contact by wearing protective clothing and equipment. Storage tanks must be vented and diked. Store drums of sodium hydroxide separate from acids, metals and explosives. Provide adequate drainage. When diluting, use agitation and add concentrated sodium hydroxide to water at a controlled rate to control heat of dilution and to avoid splattering. Do not add water to sodium hydroxide. Do not store with aluminum or magnesium. Store above 60°F (16°C) to prevent freezing.

Other Precautions: Sodium hydroxide reacts with reducing sugars such as fructose, lactose, maltose, galactose, levulose, and arabinose to form carbon monoxide. While the potential for worker exposure to carbon monoxide may be small, a potential does exist during cleaning of certain dairy and possibly other industry equipment. Carbon monoxide gas can form upon contact with food and beverage products in enclosed spaces and can cause death. Follow appropriate tank entry procedures.

Special Mixing and Handling Instructions: Considerable heat is generated when water is added to sodium hydroxide; therefore, when making solutions always add the sodium hydroxide to the water with constant stirring. The water should always be lukewarm (80° - 100° F). Never start with hot or cold water. If sodium hydroxide becomes concentrated in one area, or if added too rapidly, or if added to hot or cold water, a rapid temperature increase can result in dangerous boiling and/or spattering or may cause an immediate violent eruption.

EXPOSURE CONTROLS / PERSONAL PROTECTION

Respiratory Protection: Good industrial hygiene practices recommend that engineering controls be used to reduce environmental concentrations to the permissible exposure level. However, there are some exceptions where respirators may be used to control exposure. Respirators may be used when engineering and work practice controls are not technically feasible, when such controls are in the process of being installed, or when they fail and need to be supplemented. If the use of respirators is necessary, the only respirators permitted are those that have been approved by the Mine Safety and Health Administration or by the National Institute for Occupational Safety and Health.

Ventilation: Ventilation is not usually required for sodium hydroxide solutions. Avoid creation of mist or spray. If present wear appropriate safety clothing and provide local exhaust systems. Where carbon monoxide may be generated, special ventilation may be required.

Protective Clothing: Employees should be provided with and required to use impervious clothing, gloves, face shield (eight-inch minimum), and other appropriate protective clothing necessary to prevent any possibility of skin contact with solutions of sodium hydroxide. Materials suggested for use are natural rubber, butyl rubber, neoprene, or vinyl.

Eye Protection: Employees should be provided with and required to use dust- and splash-proof safety goggles where there is any possibility of sodium hydroxide contacting the eyes. Contact lenses should not be worn when working with this chemical.

Other Protective Clothing or Equipment: Eyewash stations and safety showers must be available in the immediate work area for emergency use.

Work/Hygienic Practices: Avoid contact with the skin and avoid breathing dust or mist. Do not eat, drink, or smoke in work area. Wash hands before eating, drinking, or using toilet facilities. Do NOT place food, coffee or other drinks in the area where dusting or splashing of solutions is possible

STABILITY AND REACTIVITY

Stability: Stable

Hazardous Polymerization: Will not occur

Conditions to Avoid: Overheating in storage accelerates corrosion.

Materials to Avoid: Contact with water, acids, flammable liquids, and organic halogen compounds, especially trichloroethylene, may cause fires and explosions. Contact with metals such as aluminum, tin, and zinc and alloys containing these metals cause formation of flammable hydrogen gas. Contact with nitromethane and other similar nitro compounds cause formation of shock-sensitive salts. Contact with water releases heat which can result in boiling and splattering. Sodium hydroxide, even in fairly dilute solution, reacts readily with various sugars to produce carbon monoxide.

Hazardous Decomposition Products: None.

TOXICOLOGICAL INFORMATION

Sodium hydroxide is a strong alkali; the mist, dust and solutions cause severe injury to the eyes, mucous membranes, and skin. Although inhalation is usually of secondary importance in industrial exposures, the effects from the dust or mist will vary from mild irritation of the nose at 2 mg/m^3 to severe pneumonitis, depending on the severity of exposure. The greatest industrial hazard is rapid tissue destruction of eyes or skin upon contact with either the solid or with concentrated solutions. Contact with the eyes causes disintegration and sloughing of conjunctival and corneal epithelium, corneal opacification, marked edema, and ulceration; after 7 to 13 days either gradual recovery begins, or there is progression of ulceration and corneal opacification. Complications of severe eye burns are symblepharon (adhesion of the lid to the eyeball) with overgrowth of the cornea by a vascularized membrane, progressive or recurrent corneal ulceration, and permanent corneal opacification. On the skin, solutions of 25 to 50% cause the sensation of irritation within about 3 minutes; with solutions of 4%, this does not occur until after several hours. If not removed from the skin, severe burns with deep ulceration will occur; exposure to the dust or mist may cause multiple small burns, with temporary loss of hair. Ingestion produces severe pain in the esophagus and stomach, corrosion of the lips, mouth, tongue, and pharynx and the vomiting of large pieces of mucosa; cases of squamous cell carcinoma of the esophagus have occurred with latent periods of 12 to 42 years after ingestion; these cancers may have been sequelae of tissue destruction and possibly scar formation rather than from a direct carcinogenic action of sodium hydroxide itself. Sodium hydroxide: irritation data: skin, rabbit: 500 mg/24H; severe; eye rabbit: 50 ug/24H severe. Investigated as a mutagen.

DISPOSAL CONSIDERATIONS

Whatever cannot be saved for recovery or recycling should be managed in an appropriate and approved waste facility. Although not a listed RCRA hazardous waste, this material may exhibit one or more characteristics of a hazardous waste and require appropriate analysis to determine specific disposal requirements. Processing, use or contamination of this product may change the waste management options. State and local disposal regulations may differ from federal disposal regulations. Dispose of container and unused contents in accordance with federal, state and local requirements. Do not flush to sewer

Index

www.ingramcontent.com/pod-product-compliance
Lightning Source LLC
Chambersburg PA
CBHW071627170526
45166CB00003B/1231